Building *Kathy Ann*

My Lake Superior Boat Project

Robert "Buzz" Johnson

Modern History Press

Ann Arbor, MI

ISBN 979-8-89656-054-8 paperback
ISBN 979-8-89656-055-5 hardcover
ISBN 979-8-89656-056-2 eBook

Published by
Modern History Press
5145 Pontiac Trail
Ann Arbor, MI 48105
tollfree 888-761-6268
info@ModernHistoryPress.com
www.ModernHistoryPress.com

Distributed by
Ingram Book Group (USA, CAN, UK, EU, AU)

Contents

Table of Figures

Introduction

Kathleen and Buzz Johnson at Christening of *Kathy Ann* at Eagle Harbor

While building *Kathy Ann*, there was nothing as uplifting as a visit from our young grandchildren. Their chatter and ability to scurry about the boat provided a welcome break from my work. I doubt that they fully comprehended what Grandpa was building. This all ended once the steering wheel was installed. The piloting started with great imagination. After launching the boat, the kids moved in and made themselves at home. Yes, this is what building the boat was all about.

Kathy Ann is a creation I enjoyed building, but in the back of my mind, the boat was always something to share with family and friends. It is a classroom to teach seamanship, navigation, and responsibility. *Kathy Ann* opens the door to a great adventure, to see sites that most people never get to see.

Kathy Ann was built with longevity in mind. Seeing "fish tugs" built of steel in the thirties and forties still working every day influenced me to build a steel hull boat. Selecting a time-proven engine was keeping in mind the rule of *KISS* (keep it simple stupid!)—one wire to the fuel solenoid, a starter circuit, and a fuel supply is all that is needed to keep the Cummins 4BT engine running.

It was a long trip from start to completion. I was still working part-time with my sons for the first two years in the family construction business and realized that I would be an Old Man by the time the boat was completed at that rate. For the next three years, the obsession of completing the boat became a full-time project. There were many nights that Kathleen kindly lifted my book and off my face and turned out the light. My wardrobe of work clothes had been depleted, with all the pieces having burnt holes, torn pockets and knees, and paint by the time the boat was completed.

Now that I had a boat to travel to the next bay and beyond, my wife's greatest fear of not seeing me during the summer came. It is not that I will not venture out, but I would like to think she would feel at home with me on the boat. Kathleen, you were most patient, and we all love you.

Dedication

My father – Elmer Johnson in his shop

This book is dedicated to my father, Elmer L. Johnson. He was more than my mentor, he was my father and best friend, and a patient teacher.

Our close relationship started with "helping" him in the cow barn at an early age or maybe it was being brought to that big fishing hole in the Traprock River to fish well after dark in hope of catching that big trout. There is a long list of occasions that brought us close together.

My father was a true craftsman. Items he made of wood were utilitarian, felt good in your hands, and were always perfect. Among his greatest crafted items were his hand carved pictures which were foreshortened 3-D cut into wide pine boards—pictures of buildings, bridges, and wildlife.

Elmer made several pairs of snowshoes, cedar baskets, and as Christmas presents, wooden toys. In later years he finally had the time and space to build traditional cedar-canvas canoes and also repair wooden canoes. Like everything he did, they were perfect from form to finish and works of art.

Most of all, Elmer guided all his children in an often unspoken way with patience and kindness. He gave us the opportunity to achieve our goals. He supported us and made us the people we are today. Our path to excel started with excellent work ethics and community obligation.

Thank you for the great start in life Dad. I think of you every time I put cabinet hinges on, trim a window, or go boating on the *Kathy Ann.*

1 My Path to Building *Kathy Ann*

Where I Began

My father's grandparents left a place in Sweden where you couldn't tell the potatoes from the rocks in the late 1800's. They settled in Scandia, in the Upper Peninsula of Michigan to be part of a community of other Swedish immigrants who worked as subsistence farmers, tradesmen, and loggers. My mother's family immigrated from Finland to the Keweenaw at the turn of the century and excelled at blacksmithing and carpentry. I spent a lot of time growing up with my mother's parents, who lived on a marginal farm a few miles away from my home—this period had a significant influence on my life.

My father, Elmer Johnson, had lasting scars from growing up during the Great Depression which influenced his very conservative way of life. He also lost his father during the depression from a heart attack. My father was called to serve during World War II in the Navy soon after graduating from high school. He was part of the crew that commissioned the destroyer, USS Newcomb. His ship was severely damaged by the Japanese off Okinawa towards the end of the war. He opted to stay with the ship for the long ride to San Francisco under tow from an ocean tug. Part of the deal was that he would get a month's leave if he stayed with the ship. His leave was spent in Detroit where his parents had moved to find better jobs as the United States went on "war footing" to produce equipment for the war.

My mother, Jean Kiilunen, had gone to Detroit to live with a relative and find work after graduating from high school. As fate would have it, on her last day with ticket in hand, before taking the train back home, my mother met my father on the eve of her departure. My father had borrowed the entertainer's accordion while he was on break and while playing, spotted Jean. She decided to stay a few more days. Marriage followed after Elmer was mustered out, and soon after that, the decision was made to move back up to my mother's home, beautiful Traprock Valley in the Keweenaw.

I was born in 1948, shortly after my parents had bought a farm which I now live on. There are indelible memories of helping my father with the dairy barn chores as far back as my memory reaches. He fixed me up with a small bucket and shovel and taught me how to take care of the calves. Getting a ride to the barn on the back of our favorite Guernsey cow at milking time was a thrill. After the barn chores were done in the winter, my father built cabinets and installed them, taking me along sometimes to spend the day with the friendly customers. His shop was crude, with a floor of boards where he worked. The rest of the floor was dirt where I played with my trucks and tractors. In those days, no one had a plowed driveway in the winter. Instead, a path from the road was packed with snowshoes. The milk cans were hauled to the road on a sleigh and the feed and supplies were hauled back to the house and barn by the same means. The cabinets were pulled to the road on a sleigh and loaded onto the roof rack on the old Pontiac by sliding them from the sleigh on the high snowbank.

Life on the farm was a Spartan existence. My parents made the move to try their hand at farming at the worst possible time. The "milkhouse laws" were going into effect and it drove many small farmers like my parents out of the dairy business. I think back at how hard my parents worked to make a living. Those few years on the farm are worthy of a book in itself.

After the cows were sold, we moved to Lake Linden where we had an indoor toilet, hot water, and a bathtub. Life was good. For the first time I had friends to play with. There were lots of cobbled-together carts to cruise down the streets and sheets of cardboard to slide down the steep hillsides on the grass in summer and skating, skiing, and snow forts in the winter. My father was away during the week on construction projects as an electrician, a trade he learned in the Navy, so I was the lawn mower in summer and path shoveler in winter.

My father had his belly full of working out of town after a few years. He started as the shop man for a local lumber yard and did side jobs in the evenings on which I often tagged along. A few years later he started carpentry on his own. We found ourselves building a new house back on the old farm when I was eleven and I was handed a vital part of the project—standing on the tailgate of the old pick-up truck pumping water from a pitcher pump for all the concrete and masonry work. So started my summers working with my father. Most kids played baseball and swam, but I enjoyed learning and working with my dad.

Strawberry Season

I had a cousin, Keith, staying with my near-by grandparents, so I was welcome to stay there often to play with him starting when I was about six or seven. My grandfather raised strawberries and shipped them to Chicago and Minncapolis with the local growers association. From mid-June to mid-July, we would be called downstairs very early in the morning as Grandma was leaving the house to set up for the day's strawberry picking. We would find our clothes laid out on the kitchen floor like a scarecrow and breakfast on the table. Carloads of kids would arrive. The start of picking found us in cold dewy rows and the buzz of annoying mosquitoes all around. As the sun rose, the day turned hot with horseflies and deerflies biting your neck and ears.

After the morning in the strawberry fields was over, great adventures began. There were great things to do in every direction. Our big adventure was hiking into the Wildcat Swamp and fishing for speckled trout. Sometimes we would roast them on a stick over a small fire. Occasionally we would pack a pup tent and provisions and spend a night. After shipping season was over, we would work in the new strawberry fields, hand weeding and setting plant runners. Sometimes, Grandpa would put a dollar bill in the row anchored by a rock to check how well we were doing our job. You better have the right answer when asked about it at the supper table.

Grandpa would take us on a memorable fishing trip just before school started—usually the day after trout season ended. We always had a good catch, but never as good as Grandpa's. We had a modern rod and real and grandpa has a long sapling with a hank of fishing line tied to it.

Blacksmith Shop

After his day job was over, Grandpa was always busy doing something. Often, we would hear *tink tink tink* coming from the old blacksmith shop. This shop had been the hub of activity for the small Finnish farm community from 1906 till modern technology and commerce rendered it obsolete. The shop still has a lingering smell of the hard coal and iron and the dirty windows from busier days. As we eased our way through the heavy wood door, we would watch Grandpa work between the forge and the anvil and wait for his slow glance to see us standing there. A short invitation was issued for one of us to crank the blower. He never told us exactly how to do it but was quick to tell us when we were doing it wrong.

Grandpa's most common blacksmith project was forging blades for a batch of Finnish knives (puukko). He would start with about six or eight pieces of well-chosen steel rod heated in the forge and repeating step after step with each rod on the anvil as they were transformed into a utilitarian ridge backed knife blade. The step I never grasped until later on was the process of tempering the blade. The heavy wood door opening was adjusted to let the right amount of light in so he could better judge the color of the steel as he pulled it out of the forge. When the color was right, he would watch the color as he drew the heat out of the thin edge by placing it on the anvil. Always watching the color of the steel, he would quench the blade in a bucket of water.

The knife blades were forged with a long tine, like a file end with a threaded end. The next step was the handle. It was always a piece of fine craftsmanship. Sometimes, he made it using an aluminum piece for the hilt and cap, or maybe it would be burl or scraps of wood saved from a custom gunstock. The center portion was usually birch bark segments sandwiched together with pieced of aluminum and copper sheet for a little color, always shaped to a form that fit your hand. The last step was to grind the blade on the big whetstone to transform the utilitarian blade into a thing of beauty. The whetstone had a heavy wood frame with a long shaft to turn the big stone which had a crank handle on the end. My cousin and I had turns at the crank as Grandpa wetted the stone from an old coffee pot and ground away, pausing now and then to inspect his work.

I took for granted all of the handmade items used as everyday tools. Clawhammer, sheep shears, copper tea kettle, carving tools, scythe blades, and puukkos. Grandma had a special long thin puukko which she used for butchering. It had been used by my great grandfather for shaving—the strop for sharpening still hung on the hall door.

Sauna Night

Saturday night was special in the Finnish community, it was sauna night. It was a time to keep connected with those who worked with one another as an informal coop. I had a few favorite homes we would visit with my grandparents, but the commercial fishermen at Big Travers Bay (in Keweenaw County) topped the list. We usually would arrive in the late afternoon in summer so my cousin and I could brave a dip in the cold water of the Big Lake or drown a few worms in the river.

You could hear the drone of the diesel engine from the returning fish tug long before you could see it. First, you would see a smudge on the horizon; this was the cloud of seagulls scavenging as the men dressed their catch on the trip home. Next, you would see the white of the bow wave and soon the boat would materialize on the horizon. Once the boat was tied up, the men would efficiently store the catch in the icehouse and prepare the boat for the next trip. This "fish tug" was the Atomic, a new boat that my father and grandfather had a hand in fitting out.

Next to the Atomic was a big graceful green and white fish tug named the *Twin Disc*. I couldn't figure out why it was called the *Twin Disc* when you could only see one propeller. It had originally been built for the Twin Disc Company as a test bed for their transmissions in the 30's. Stories by the fishermen made the picture of the Twin Disc clearer.

As the evening fell and the sauna got hot, it was the kid's time for the sauna. After a little time in the heat of the sauna, Grandma would appear to make sure we had cleaned behind our ears, got us dressed, and readied the sauna for the next shift. Sauna was a communal affair so the women would go as one group and the men as another. When the women went for their sauna, the table was set for the men to have coffee and bakery. This was the best part. The stories started. Often it was a repeat of the same story told by another fisherman. There was much discussion over the new Loran-C, steel boats, diesel engines, Twin Disc transmissions, and the amazing autopilot. The stories ranged from tragedy to great catches and were often told with great animation. There was always a spot at the table for us kids.

Commercial fishing on the Great Lakes was subsidized during World War II, so these fishermen had the money to be well equipped and productive. The fishing boom lasted until the mid-50's. The arrival of the invasive sea lamprey, reworking of commercial fishing regulations, and overfishing brought most viable commercial fishing to an end.

Soap Box Derby

1963 World Championship, Soap Box Derby - Winner of first Heat

Soap Box Derby, "The Gravity Grand Prix" was a popular when I was an early teen. It took some convincing, but my father relented and agreed to work with me to build a car. The wheels and axles came from the local Chevy Dealer and our friendly banker contributed some money for material in exchange for painting their name on the car. It started with drawing a silhouette on cardboard of me in the driving position and designing the car around me. One day, my dad came home with an oak library table that had been left out in the weather.

By the time the car was finished, most of the table was built into the car. I probably weighed about 90 pounds and the maximum total weight for car and driver was 250 pounds, so I had a lot of weight to build into the car. Grandpa took the unfinished car to my uncle's repair shop one Saturday and custom fit the steering wheel with me sitting in the car. I spent about 650 hours building the car. It was a good-looking creation, good enough to win the Hancock Soap Box Derby for 1963. From there, on to Akron, Ohio with my family for the world championship. What an eye-opening experience for a kid. I won my first heat at Akron. The seed was planted for what could be accomplished with a little effort.

Gunstock

I purchased a single shot shotgun with a broken stock held together with electrical tape when I was old enough to start hunting. One day my grandpa showed up at my dad's shop with a chunk of yellow birch and worked with me to fit the chunk of wood to the gun's action. He handed it to me with a short directive. "Finish it." Grandpa always had a project in the works. He made an ice fishing sleigh one winter and always had a gunsmithing project or two in the works. One day while he was working on the handle of a puukko, he noticed that it had a cracked hilt piece. He handed it to me to complete.

College and Beyond

After high school, I decided to take a course in architectural drawing at what was then Ferris State College. At the time of graduation, I faced a dilemma—should I head toward the city to earn big bucks or go back to my roots and live the life I loved. Although I never pursued a job as an architectural draftsman, the education I received I use every day. Ferris may not have taught some of the areas of construction in great detail, but it opened the door, so I knew where to find the answer. Term breaks and the time after graduation had my father waiting to put me to work.

I finished college in March 1968. I had the summer to work and enjoy the great outdoors before I was drafted into the Army. After basic training I was handed orders which I did not completely understand. All I knew was I was to report to Fort Hood, Texas. I soon found myself in the Orderly Room of C Company, 63rd Engineers. I learned that I was classified as a Carpenter, but all of the carpenter slots were filled so I was placed as the company draftsman. A few weeks later, I found myself in the position of Construction Records Clerk and spent my time dealing with green Second Lieutenants straight out of OCS. I would have a few good stories to tell about that adventure.

On to Vietnam. From Texas, I received orders to go to Vietnam, again as a carpenter. I was dropped off at the headquarters of 169th Engineers and never made it out of Headquarters Office. I was offered the job as Battalion Morning Report Clerk. In Vietnam, we took part in rebuilding their highway system. The young equipment operators did a great job. On leave before going to Vietnam, I spent time ice fishing and took some interesting pictures while on the ice. I would show them to the Vietnamese. They could not grasp that I was walking on the top of the water. Experiencing life in the military has instilled a quality of life that I carry with me every day.

My job was waiting when I came home from the Army, and I soon was back to work. It was early January. On the first day at work, it was about thirty degrees below zero, and I ended up in bed for ten days with swollen glands from the extreme climate change. I worked in the residential field with my father and finally, with my own company for a few years. It was during this time that Bob Carlton who had worked with my father in the early 50's wanted to build a retirement home at Agate Harbor in the Keweenaw. After a day's work on Bob's house, I decided to stop and visit his eldest daughter, Kathleen on the way home. A year later, we were married. Kathleen's old uncle Jack joked that I was given Bob's eldest daughter as the down payment on his house.

We had some bumpy days in our young business those first years of marriage, so I decided to step back and head in another direction. I went to work in a management position for a larger construction company. I was the guy in the office with blue jeans and leather boots so I was handed the projects that could not be estimated in cubic yards, units, or square feet. I enjoyed the challenge and learned a lot. I often went into the field to run the job if we got the contract. Part of my job was analyzing problems and errors and writing reports on what caused the problem and how to fix it. My philosophy was, "You can't fix it until you find out why it broke."

One day, after thirteen years, "the light came on"; I was missing the opportunity of working with my young sons in our own company. REJ Contracting was born. It wasn't long before my old customer base returned. We built custom homes in Keweenaw and the surrounding area along with a few commercial projects. Then came the recession of 2008. There was no work to be had in Keweenaw, so we took whatever was offered when the phone rang. Those were some skinny years. As work picked up after the recession we decided to diversify and found ourselves doing more restoration and repair of old buildings and working on lighthouses around the Great Lakes.

Along the way through my years in construction, I had the good fortune of working alongside some very skilled people in other trades. My neighbor, Al Koskela, was a mechanical contractor that I worked for now and then. He had his sheet metal shop in the basement of my carpenter shop for several years. My father-in-law, Bob Carlton was a master electrician who was often recruited to wire projects for the family. He had the patience to teach us the "hows and whys" of installation as he worked. My involvement as a design-build contractor also taught me about other trades. As subcontractors, other tradesmen were often involved with the design of projects as well as the installation for their trade.

REJ Contracting has grown from a company with a few employees doing residential building to a company with a diversity of talents backed by a depth of knowledge and experience and well equipped to handle most any task.

As I approached 70 years of age, I backed away from the company business a bit and the last few years, seeing that the boys were managing to get along without me, very well without me in fact, I had to put the big push on finishing the *Kathy Ann* so I could enjoy her for a few years.

2 In the Beginning...

I grew up on the Keweenaw Peninsula, a spine of rocky land jutting out almost to the center of Lake Superior. Hunting, fishing, and boating are a function of most everyone's life here. We have plenty of inland lakes for water sports and fishing but Lake Superior, the "Big Lake" is the "Big League" for local boaters and sportsman. When there are gentle waves wetting the pebbles and sand on the beach, she is beautiful. When she is in a foul mood, she can be awesome and deadly. Sooner or later, those who venture out onto Lake Superior have a story to tell. The lure of the Big Lake for adventure and fishing causes a stirring inside of most boaters which leads to the yearning to get a bigger boat—this is called "Two-Foot-Itis".

So, with the apprehensive approval of family members, you start shopping for that bigger boat. Shopping for the next boat should not be a decision made on a whim but conducted as a study of what you need and what is available to fill this need and what costs are involved. Remember, the cheaper the boat, the more work and cost involved in making it seaworthy. If you are not familiar with boats or are not mechanically inclined, invite someone along with you when shopping for a boat who is. Finding the right boat may be as simple as spotting one with a For Sale sign in a front yard, slamming on the brakes and taking a look or as intensive as an exhausting on-line search.

Now that we have you thinking about the next boat, we shall add another part to the equation—have you thought about building your next boat? Some of us have the equipment and the facility to build a boat. Some have the experience and skills to build a boat. Some have the time and drive to build a boat. It takes all of the above elements, or you well could have a monument forever in your backyard. Your first boat project need not be a big one. Much satisfaction can be had by starting small, such as building a sailing dingy or a cedar strip canoe or just fixing up your present boat. The experience gained during your first boat project would be a good gauge of your abilities to build something larger. If you decide to build a bigger boat, much enjoyment can be had in the meantime with your small creation.

Our first boat was a twelve-foot plywood boat which I helped my father build. The choice of red oak for framing lead to an early demise because it was not very rot resistant. She was quite seaworthy for its length and could handle two adults and a few kids. She could be manhandled onto a home-built trailer which made exploring small lakes a lot of fun. The little boat had been built in the old barn where the part can of left-over green paint sat unnoticed for years, until one day my youngest brother and a few of his neighbor friends accompanied by their Colley dog opened it up. A phone call from the neighbor around dinner time that evening stating that, "Our dog came home painted green, and it came from your direction." When we sat down for dinner, Dad started the interrogation. The answer was soon to come. "We found this can of paint and an old brush and the dog kept walking by."

One day my father came home with a fourteen-foot Cruisers Runabout, a great deal better than the plywood boat. She sat on a legitimate boat trailer which made life easy. We made many fishing trips on the Big Lake (Lake Superior). She was a fair-weather boat; however, you would sometimes take water over the transom when trolling with a following sea. She was a beautiful wood lapstrake craft.

A seventeen-foot Thompson Tomboy came next. She was an excellent fishing boat for the Big Lake. We made a few trips to Isle Royale with her placed on the deck of the Ranger III. She was a plywood lapstrake craft and is still in the family and is in beautiful condition. One beautiful afternoon my dad and I were trolling off Bear Bluff in our Tomboy. With us was my airplane mechanic friend Mike. We were watching a Loon swimming nearby. The Loon decided to take off. In its awkward paddling, running, flapping display it finally took flight. Dad looked at Mike and asked, "Did you work on that one?" Ah! You would have had to have been there to appreciate the joke.

The pace of married life, four kids, and work quickened. It was a while before another boat came into my life. This was a twenty-five-foot lapstrake Carver with twin Chevy four-cylinder engines. She was purchased for transportation to Isle Royale for a construction project I was involved with. It was about a fifty-five-mile open water passage from Agate Harbor. Those little engines purred but she was rather slow. She came with a vintage Loran set which was about the size of a typewriter. You had to plot your destination on a chart that had the loran signal overlay on it and punch the numbers into the machine.

Forty passages were made that summer to and from Isle Royale as well as many fishing excursions for employee therapy and shuttles between Mott Island and Snug Harbor on Isle Royale. One mid-September afternoon on our fortieth trip traveling home for the weekend, we were amongst those long period waves from the Southwest that late summer brings to Lake Superior. After slowing down to ride over the crest of a big wave I couldn't get the boat back on plane. We discovered that the boat was filling with water from a source we could not find.

I had one of the crew aim for Eagle Harbor, which I determined by looking at the profile of the Keweenaw Peninsula while I searched for the source of the leak. I lifted the cushion covering the engines and saw the water almost up to the distributors, put the cushion back in place and said, "Start bailing." A mayday call was made to the Portage Coast Station for some assistance. We were in the middle of Lake Superior. After all hands baling for about an hour and a half, the Coast Guard arrived in force. A thirty-foot work boat came from Dollar Bay, a helicopter from Traverse City, and a Boston Whaler that had been trailered up to Eagle Harbor all arrived at about the same time. The Coast Guard managed to get a crew member aboard from the Boston Whaler and high lined over a pump which enabled us to wallow our way into Eagle Harbor and on to the boat trailer.

This was not the end of her. We uncovered the history of what caused the broken planking. She had been run onto the rocks at Copper Harbor in the fog years earlier while returning from Isle Royale. We sistered in a few ribs and replaced a few strakes, tightened up the hundreds of brass bolts on the seams and gave her a good paint job. We ran the old girl for a few more years before two-foot-itis struck again.

The purchase of a twenty-eight-foot Chris Craft Catalina came next. It was an exhilarating moment. She gave us room to move, a few conveniences, and no wood to maintain. We added a fifty-gallon fuel tank to increase our cruising range and installed good electronics. We eventually had to replace the port engine. Along with being an excellent fishing boat, she served as our floating pick-up truck for working on remote lighthouse repair projects and construction projects on Isle Royale.

One memorable excursion was voyaging to and from a mid-September project at Windigo on the west end of Isle Royale. Late summer on Lake Superior brings long periods of southwest wind and growing long period waves. We waited for a break in the wind. Finally, a wind shift was in the forecast, so I took my son Joe out of school and moved the boat up to North Portage Entry and listened to the NOAA weather forecast as it updated every hour. We waited as late in the day as we could and finally made the decision to try it. The wind shift never came. Everything loose was put on the floor, and we ventured out into the cross sea which rolled the boat but was not bouncy.

After seventy neck tensing miles we make the turn into Grace Harbor with only enough daylight to see the silhouettes of the hills on each side. As luck would have it, there was a camper with a Coleman lantern hanging at Grace Island dock which we used as a guide. The approach to the dock was not very graceful but we had made it. The next morning, we made the short run to Windigo. When we completed the job, the trip home was not much better. The long-range forecast did not call for any significant change in the weather, so we ran back in the same conditions.

The big jump in gasoline cost dampened our enjoyment of making extended fishing trips. We did, however, put her to work on a few remote lighthouse repair jobs and Isle Royale projects. The Catalina sat in the shop collecting dust for a few years "just in case" we had a job for her. Well, we did find one more job for her. We were contracted to do repair work on the Isle Royal Lighthouse on Menagerie Island. The lighthouse was located on this small island at about the center of the south side of Isle Royale. It was a 110-mile round trip from Agate Harbor which was stretching the safe limits of our 145-gallon fuel capacity. Other fueling options for this run did not work out much better as the closest fuel on Isle Royale was twenty-five miles away. The first trip out with supplies left us searching for the route the lighthouse keepers used to get up the rock face from our ten-foot RIB. There was only a small patch of boulders close to the island where an anchor could be set with only a short scope of anchor rode, so we had to keep one eye on our boat.

We spent the night at Malone Bay where there was a small visitors center with a picture of the lighthouse we could study, searching for the original access route. The second trip with supplies left us searching again for the route up the rocks and again we looked at the picture as we spent the night at Malone Bay. On the third trip we brought out the crew to start work and bring provisions for our shelters at Malone Bay. Travel to and from the lighthouse was a two-mile run with our ten-foot RIB. There was not a trip made that we did not get wet. After further study of the lighthouse picture, we realized that whoever reproduced the picture had the negative backwards. It didn't make any difference because the steps were long gone.

The project was completed in early October. We hurried to load the boat for the trip back because a front was forecast to move through in the afternoon. We didn't make the weather window so had to spend the night. The plan was to leave at 5:00 a.m. because another front was predicted to move through late morning. I stepped out of the boat onto the dock in the dark and found myself sliding on my back on the frosty dock headed for the drink but managed to stop short. We left in the dark following our GPS track and arrived at Agate Harbor just as the sea was building. We had to make one more trip to Isle Royale Lighthouse in early November with the Government Inspector. I made sure the inspector understood that I would pick the day. The day was bright, clear, and relatively calm and every breath could be seen in the cold air. Lake Superior is a lonely place that late in the year.

The trusty Catalina was returned to the shop to collect dust. As the *Kathy Ann* was taking shape, the decision was made to sell her. The Catalina is now in new hands that will give her the TLC she needs.

The twinkle of an idea for a boat which became the *Kathy Ann* started years ago. The dream morphed as time progressed from finding an old steel "fish tug" and fixing it up, to acquiring a used trawler, many of which were available in the aftermath of the 2008 recession. Finally, I realized that I had the ability to build a trawler from scratch rather than buy a fixer-upper. With time, I progressed with the study and search for a length and design of a trawler to fit my desires. This part of the project was an education unto itself. For Christmas the family presented me with books on boat design and calculations for various aspects of design and what equipment I would need. Getting a membership to the American Boat and Yacht Counsel (ABYC) with its catalog of standards was the best decision I made while planning to build the boat and was always at hand thereafter.

After much research, all the desired design ingredients as I uncovered them were thrown in the pot and stirred. I searched for the design of the best boat to fit our need, a boat designed by Dudley Dix, a trawler named *Echo 38* rose to the top of the pot (dixdesign.com). A trawler is a displacement hull boat with a hull speed determined by its length. A short list of features we were looking for:

- Steel hull
- Propeller protected by a rudder shoe
- A length between 36' to 40' to achieve an acceptable efficient hull speed
- Enclosed accommodation for passages on cold Lake Superior
- Easily accessible engine space
- Tankage sufficient for long range cruising
- Raised Pilothouse were much of our time will be spent
- Fuel efficiency

Now the die was cast for where we were headed. Now we could start the boat project. It was a little like putting your foot in Lake Superior and then getting up the nerve to get the rest of the way into that very cold water.

I had been collecting parts for boat construction for quite a while. I came across a big standing dead red oak while cutting firewood with a perfect sweep in the trunk for pilot house roof beams. This was sawed up on our portable mill and put to dry for a few years before the boat was started. White pine and cedar wainscot was made from slabs I collected when the boys cut lumber on the portable mill. I dried and milled the wainscot in the carpenter shop. The cherry paneling was from a house I was asked to finish off and the owner had to get the unused paneling out of the house. Then there was the steel salvaged over time that

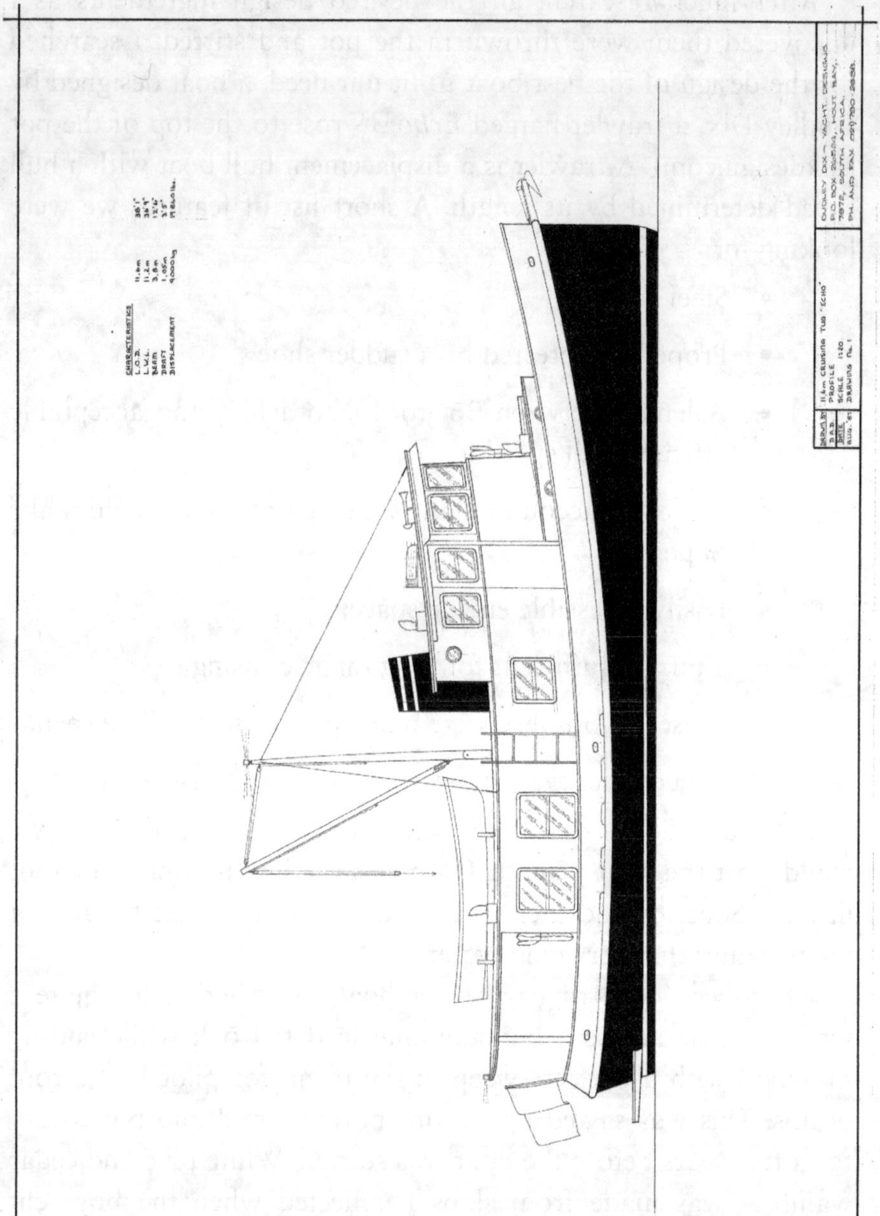

***Echo 38* Drawing #1, by Dudley Dix**

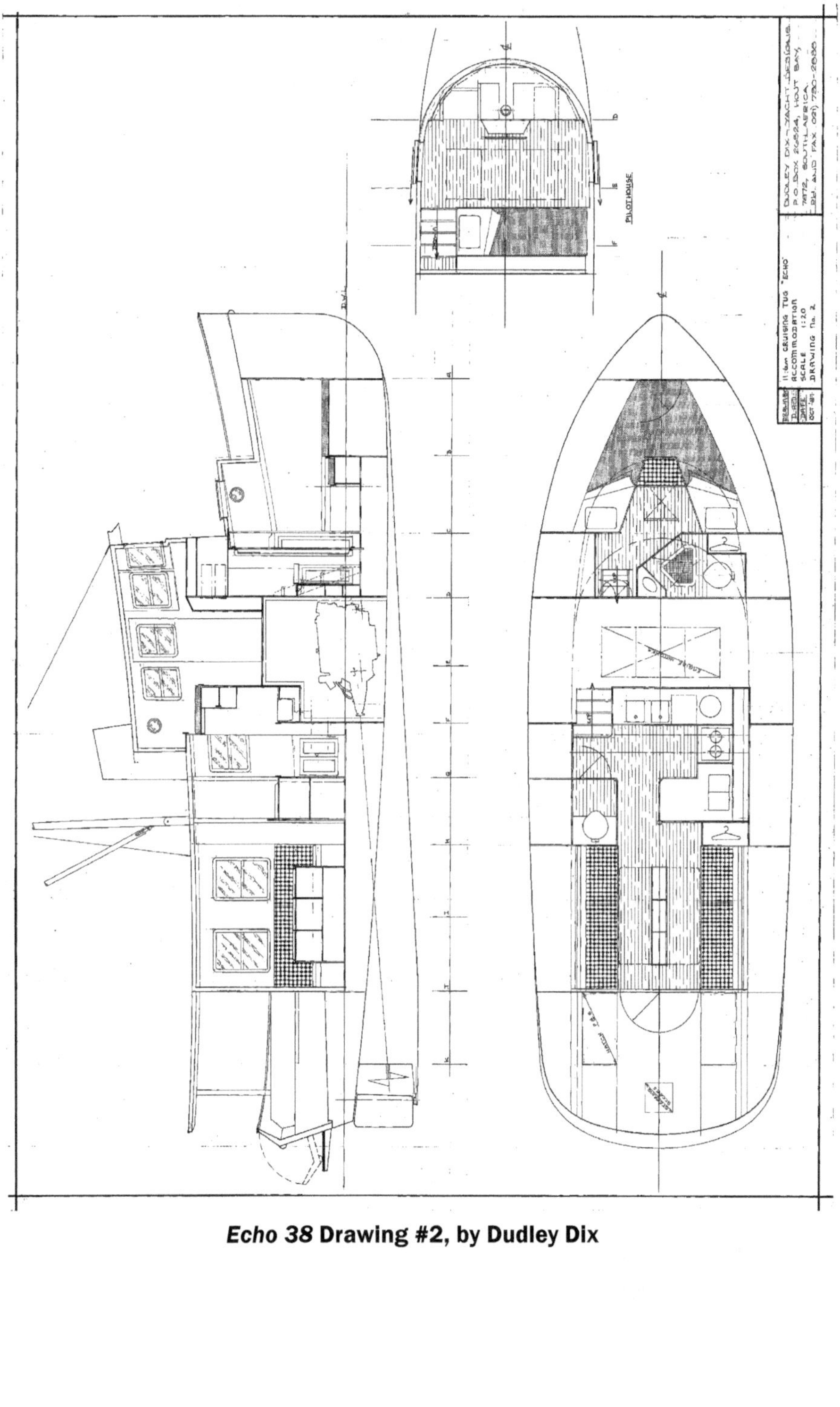

***Echo 38* Drawing #2, by Dudley Dix**

was rummaged through often and a collection of aluminum pieces which were put to good use.

A family friend, Andy, gave us an old RV that a big pine tree fell on and thought it could be repaired but we found it was beyond its prime, so we parted it out, saving items such as appliances and windows with thoughts of using them in the boat. The only item we repurposed in the end was the kitchen stove and the medicine cabinet. Oh, that Andy was a sly fox—he knew where that pile of junk would end up.

Slowly, the wheels were set in motion to sell the idea of building a boat to my wife and family. Kathleen was thinking "rowboat", Rob, my oldest son, couldn't understand why anyone would ever spend time building a boat, and Joe who loved boats was far too busy.

My father-in law, Bob Carlton, who was a Navy veteran of three wars, upon hearing about the project made sure I understood proper nomenclature—decks, bulkheads, and overhead. He was always eager to get a progress update on the boat. Bob had navy in his veins. We made a trip with our kids when they were young to Gulf Shores, Alabama where Bob and his wife Peg were wintering. Finally, Bob had visitors that would enjoy a tour of the Battleship Alabama up in Mobile. Winding our way through the bowels of the ship with Bob was like having our own tour guide. Bob was giving us a very animated explanation of what his responsibilities were for "damage control". Observing the boys, I noticed the signs that it was time to look for a head (bathroom for you landlubbers). I had a hard time getting Bob's attention so he could take a break. It was a great day. The boys slept all the way back to Gulf Shores.

Kathleen may have been thinking "rowboat", but I had been thinking "done in two years". I realize now that even if I knew what I was doing, I would have doubled my time estimate, and still would have been off on my estimate by a few years on finishing the boat.

3 Big Step—From Vision to Reality

It was winter 2017. I was about to embark on a project my dear wife did not comprehend. The drawings for the *Echo 38* had arrived from Dudley Dix. Since I had a background in architectural drawing, I understood the drawings quite well. The more I looked, the more information I uncovered. It was a bit like learning a new language. The decision of where and how to start building the boat had to be thought out. Our production carpenter shop in the basement of an old church building in Calumet had enough floor space to loft the entire length cross-section, an area of about forty feet by twelve feet with room to walk around, and an overlay of half hull plan full scale. I shuffled a few benches and tools around to clear the area to set up the lofting table. "Lofting" is an old nautical term which referred to the loft or attic of the boat shop—the only free space large enough to draw out patterns for their boats.

The purchase of sheets of stable fir plywood painted white set up on short sawhorses for the lofting table made life easy. A few hard pencils, fine line markers of various colors, small slim nails, straight edges, and cedar battens were the only tools needed to convert the table of offsets and plans into full sized patterns. The table of offsets which was part of the plans provides a plot of points along the hull frames, measured from the vertical center line and the horizontal reference line which for this boat is the assumed waterline.

Accuracy at this point was critical. All the frames were drawn out with different colors using the same center line and reference horizontal line. Errors were easy to spot, and a sense of the flow of the shape came to life. Years of experience showed me how minor errors can compound as the project progresses. Thin plywood patterns were made for frame parts to fit the lofted layout as well as tank sides, keel sides, stern framing, bow framing, and bow radius scantlings.

Templates used as guides for cutting with the plasma cutter for common radii were also made. The plasma cutting templates were 18-gauge steel with quarter-inch standoff spacer blocks glued to the back to minimize gouging of the edges of the steel when in use with the plasma cutter.

Hours upon hours were spent out of the house with little to show Kathleen. After all, she wanted me to sit and talk with her—this is something she was very good at. After a winter of evenings and Saturdays spent hiding out at our old church carpenter shop, all I brought home was an armful of odd-shaped plywood pieces.

Spring was coming so it was time to put the carpenter shop back into commission. The layout had to be dismantled. Wait! There was one more project before the layout was disassembled. I made a mock-up of the connection between the steel hull and the wood cabin while accurate dimensions were available. The portion of the layout for the frames, bow, and transom were brought down to our mechanic shop and set up under the lean-to so parts could be fit up and tack welded together for an exact fit. Again, we are reminded that accuracy is very important as errors at this stage compound as work progresses. This was followed with a summer of converting the pieces from plywood to steel shapes of the same design.

Before construction started, we took the time to set up a workstation with two 120-volt circuits for hand tools and lighting. Two 240-volt circuits for the welder and the plasma cutter. Compressed air was also piped in from the shop compressor for minor sandblasting, air testing for leaks, operating the die grinder and plasma cutter, and clean-up. I set up our cold-cut

Checking fit of parts on Lofting Table

Cabin Framing Attached to Deck Mock-up

Pilot House Framing attached to Deck Mock-up

saw in the mechanic shop with a table to support long stock.

Along with studying the drawings came a list of decisions that had to be made on the selection of equipment such as running gear, rudder shaft stuffing box, shaft seal, prop shaft diameter, stern tube diameter, cutlass bearings, and a pile of other items which were to be incorporated into the hull of the boat. These decisions had to be made now! Hours upon hours were spent researching and selecting the correct parts of high quality. Thanks to my book library and the internet, most decisions made have worked out as anticipated. The Dudley Dix plans only covered construction of the hull in detail—the rest was up to me.

Deciding on power and gearing for the boat was a very important decision. After a study of what horsepower was needed for the Echo 38 cruising at optimum hull speed, which was about 8-½ knots or about 9-½ miles per hour, I got lucky. Our diesel mechanic neighbor, Mark, had an old bread truck with a Cummins 4BT by his shop that had sat for about fifteen years. The find was a reliable low-tech engine that, when rebuilt, could be expected to last many years and was a perfect fit for the optimum power curve. I don't know how the engine was put into the van originally but it sure was a project getting it out.

I built an engine stand at the mechanic shop for the rebuild project. The engine was pulled apart, inspected, and measured up for rebuild parts. It was discovered that a blown head gasket had made a shallow groove between two cylinders in the block, so I brought the block to an engine shop to have the deck shaved. The head was also brought along and found to have cracks between the injector holes and the valve ports. By the time I finished the rebuild, I had replaced the head, camshaft, and installed a complete rebuild kit. After all the hours the engine had been run, you could still see the honing marks in the cylinders.

I had decided on a Twin Disc MG 5050 SC transmission with 2:1 gears. I had a trolling valve and a PTO added to the transmission. The gear ratio of the transmission and the propeller diameter and pitch go hand in hand to fit the most efficient hull speed to the RPM with maximum torque of the engine. Now we had to find the proper bellhousing, damper plate, and flywheel to mate it to the engine.

After mounting the new bellhousing, I realized that the starter was located in a different position than on the original bellhousing and would conflict with the original exhaust routing and location of the turbocharger. The original configuration had the exhaust dumping out of the rear bottom of the manifold. This would not work, so the hunt was on for a manifold with the outlet on top. Many knock-off manifolds were available, but experience told me to steer clear of them. I finally found an original, never used manifold, made in England. The new manifold meant I had to reconfigure the air supply between the turbo and the air box, modify the turbo oil return line, and find a longer oil supply line for the turbo.

Finally, it was ready for the test run. I built a stand, rigged up cooling water, and ran the engine for some time before installing it in the boat.

Selecting equipment to start the steel work required some discussion here at the shop. I wanted welding equipment suitable for work on the steel boat hull that could be used in the shop or on the job later. We had a good portable stick welder, a wire welder, and a shop stick welder, but thinking about how much this tool would be used and how often it would have to be moved, the decision was made to get a good suitcase-type welder. Knowing that a lot of the welding would be done with the 1/8th-inch rod at a relatively low amperage, it was decided to get high-frequency start, a real pain-in-the-neck saver. We also made up a heavy-duty extension cord to operate the welder closer to the work so the welder lead length could be short to minimize voltage drop. You may ask "Why not use a wire welder for welding the thinner material?" It is a matter of experience and choice. Wire welding does a great looking job, but I question the ability to penetrate the joint. A wire welder also does not work well in the wind.

A plasma cutter was essential because it was fast, accurate, used only a few consumables, and, most importantly, minimized distortion with its low heat transfer to the workpiece. We already had this tool, but I cooked it halfway through the project. A replacement plasma cutter was selected with input from our local welding supply shop and proved to be a great step forward.

The cold-cut saw was a great tool. It could cut as accurately as the carpenter shop miter saw. The only problem was that inexperienced employees occasionally used it and damaged the blade.

Angle grinders are tools used often and will have a rough life. At least two grinders available at most times is handy, one grinder with a quarter-inch wheel and one with a thin cut-off wheel. The four-and-a-half-inch size works well. The paddle switch style is what I prefer as it will stop if you drop it or put it down. I looked for deals on grinding wheel packs because we went through a bunch. A die grinder would also fit in well with the angle grinder to get into corners and other hard-to-reach places when chasing leaks.

Regarding sandblasting, you may consider bringing in someone with the right equipment and a big enough compressor. In our case, we had the compressor and would find use for a good sandblasting set-up in the future. I selected a system with a three-bag hopper and a helmet with fresh air supply. Medium grit crushed slag abrasive was used and I used over a pallet of abrasive.

Along the route to completion, I collected a bunch of hole saws for cutting steel and wood, drill bits, a good battery drill and impactor set, handmade pry bars and benders, and a set of improvised bending shoes for the wood splitter to shape scantlings and rail caps.

Rigging equipment such as a coffin hoist, cable chokers, and an assortment of clamps was essential for handling material and putting working pieces into place. Access to an extended boom forklift made most tasks a one-man job.

Lastly, there was the reliable acetylene torch which was essential for cutting thick steel, bending steel, and heat shrinking steel here and there to get a fair fit-up.

There were also a lot of hand tools and jigs which were accumulated through the course of the project such as a jig to align scantlings for butt welding and an ark template for the deck and cabin top framing. I also made a few templates for cutting slots in the frames for the scantlings which also incorporated the all-important limber hole. Limber holes are openings in the hull framing to prevent water from collecting in low spots.

That is what it took to get the boat started. There is a lot of sketching, building mock-ups, and collecting more equipment to help turn thoughts into reality. Planning ahead for the placement of equipment, through-hulls, and cabin layout is critical as you proceed with the building of the hull.

As construction wound down, spare bolts, screws, electrical connectors, and other parts were put aside as part of the boat inventory. A good set of mechanics tools was purchased and a few task-specific tools were rounded up. In retrospect, the mechanics tools probably should have been the first tools to be purchased instead of the last.

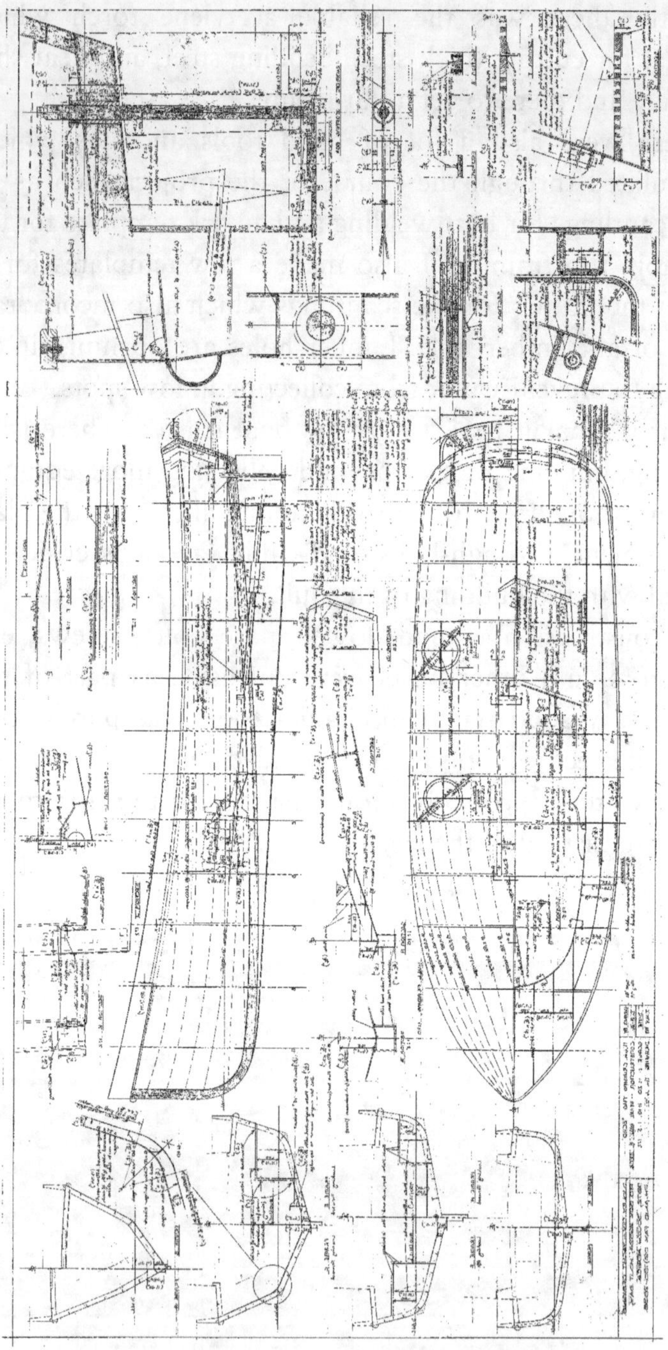

***Echo* 38 Drawing #4, by Dudley Dix**

4 A Steel Hull

The space at the back of the mechanics shop was selected for assembling the boat. The plan was to place the boat so a permanent roof could be erected over her once the hull was turned over. The area was graded, compacted and given a six-inch layer of crushed stone to provide a solid base for building. Salvaged 8"x 8" timbers were places thwartship at about four-foot on center upon which steel beams were placed so the support work, as built, would fit inside of the boat. Pipe legs of a convenient height, so I could both walk under and reach most of the hull were welded to the beams and capped with a 2"x 2" angle. The cap angle was at the reference line used to lay out the frames. I soon added a runway of plywood on top of the wood timbers to make access easier, but it reduced the headroom under the 2"x 2" angles by eight inches. Ugh! Lots of bumps and cuts on the head.

Common A36 steel was selected for building the boat. The decision was driven by cost and availability. In researching material for steel boats, I saw that pre-cleaned and primed steel plating was available at some sources, but this had mixed reviews. I also saw that Corten steel was used for plating but looked like it came with its own set of problems.

The first project was construction of the deep keel while I had a clear level spot to work. This was set up on one of one of the beams which would later support the boat. I decided to spread the top of the center portion of the keel, deviating from the drawings to provide room for a keel-cooler tank at the stern which had the stern tube running through it. The stern tube houses the propeller shaft. A day tank of about forty-five-gallon capacity for engine fuel was partitioned off at mid ship. A space between the day tank and keel cooler made a perfect spot for one bilge pump and the propeller shaft seal, and the low spot in the keel ahead of the day tank made a good spot for a second bilge pump. I had made patterns of the keel sides from the lofting table which proved to be quite accurate. Welding liquid tight joints at inside corners of tankage proved to be a frustrating experience. Good old 6011 welding rod saved the day. The keel finally passed the leak test. Compressed air blown from the inside and soapy water sprayed on the outside of the welded joints was the procedure for finding leaks. The finished keel was set aside.

There is a formula for designing the keel cooler for how many square feet of hull surface is required for the horsepower of the engine to adequately cool the engine coolant—vertical surfaces work more effectively. There is also a formula for sizing the coolant header tank for the estimated volume of coolant in the engine cooling system.

Assembling the frames went well. The pieces were placed on the pattern table with copper sheet scraps under the joints to prevent burning the pattern table. Welding parts together was a constant fight to prevent distortion. A piece of 2" x 2" angle was welded across the frame at the horizontal reference line which would be removed later. I rigged up an electric hoist on a trolley to help move the awkward sized frames.

On our erection frame, a centerline string was set up. The frame spacing was marked out and the frames tacked in place. At this point it started to look like a boat-shaped whale skeleton.

The keel was now fitted to the frames. I discovered that the keel had a slight curve to it, a big rookie mistake. It took some time to straighten it out. The sides of the keel extended into the hull so the hull plating would butt up to it. This detail also enabled me to make the day tank top level and access to the keel as simple as possible.

Hull Steel Framework Completed

Bow View of Hull Plating in Progress

At this point I concluded that there was no "convenient" height for the pipe legs as I had a daily collection of knots on my head and still had to use a short step ladder to reach the keel.

With snow, the end of outside work for 2017 came. I gave it a try to work in the snow, pulling the welder along in the snow scoop and losing tools in the snow. The worst part was slipping and sliding as I worked and realizing the risk was not worth the gain. It was time to retreat to the warm carpenter shop to work on the cabin.

Spring brought long days and a better attitude toward working on the hull again.

With the frames braced in place, installing the scantlings started. The scantlings, which are flat stock strips, were set into the slots cut out of the frames. The scantlings along with the frames defined the ultimate shape of the hull. Fitting the scantlings was a learning experience. I soon found out that there were only a few places where the scantlings fit square to the frames so most of the frame slots required some rework with the grinder. Also, pulling too hard on the scantlings to get them into place made sharp bends at the frames. I made a bender similar to a ridged conduit hand bender to prebend the scantlings ahead of easing them into the frame slots. The bow scantling pieces were cut from plate derived from patterns lifted off the lofting table because they were too sharp of a curve to be bent. Small half circles were cut into the scantlings for drainage at the low point between each frame so water would not collect when the boat was upright. Half circles were also cut where each plate joint crossed the scantling so the plate joint would a have a continuous weld.

The hull plating is three-sixteenths-inch thick from the keel to the top of the turn of the bilge. All plating above this line is one-eight-inch plate. To fit the eighteen-inch radius of the rounded chine, I had the steel suppler roll a few 4 x 8 plates to this radius and then I cut segments from them for the radius chine. The stem piece was shaped like a segment of a cone. I laid this part out by dividing it into small segments and figuring out the degree of bend at each segment. I brought it to a local fabrication shop to bend on their break. They did an accurate job but had trouble at the tight bottom end because it would not fit in the break properly. A little hand work at our shop corrected the shape and it fit perfectly.

The Tugboat style stern was an interesting part to frame up. All the information was on the plans... somewhere. Duplicating the bends of the wraparound scantlings to ensure symmetry was a project in itself. Installing the perimeter bearing plate for the cockpit sole with a drain trough across the stern required a close study of the drawings; again, the information was there... somewhere.

The bottom plating started by making a pattern of thin plywood and transferring this to the twenty-foot-long sheets of steel beginning near the middle of the hull and working toward the ends, then trimming the leading edge of the steel to fit. The free edge was left a little long so it could be trimmed in place. I then found the tangent points on each frame. This was the point where the radius of the chine transitioned to the straight on the bottom. It happened to be the welding joint made when the frames were fabricated. These points were then connected by use of a cedar batten to form a gentle arc from about the second frame to the stern.

When fit-up was acceptable, the plate was tack welded to the frames and the scantlings and tacked at about six-inch on center where the plating edges met. I discovered that when the wider plates were eased into place developed a cup in the transverse direction which required the scantlings to be pushed out to meet the plating or pressure put on the plate to push it onto the scantlings. A space of about one-sixteenth-of-an-inch between plates was made to ensure complete penetration of the weld by cutting in between the tack welds with a cut-off wheel where necessary.

Now came the challenge of fitting up the radius chine pieces. These plates had a compound curve, meaning that they are not only segments of the eighteen-inch radius plate but also must follow the curve of the length of the hull. Cutting these pieces with an acetylene torch helped by causing some shrinkage distortion at the edges which was in my favor. Where the curve was most pronounced, shorter segments of plate were used to make a fairer hull. Still, some heat shrinking was necessary along the edges for an acceptable fit.

I found it handy to make up a few dogs which were welded to one side of the plating joint and wedged with a cold chisel or other wedge to bring joints into alignment. A quick lesson learned was to weld the dog only on one side with a small tack weld on the other side so it could be removed with a few blows of a hammer. The thin plywood patterns were reused multiple times, getting smaller and smaller each time. A rule to follow, never make a plating joint over a frame or scantling because you cannot get a complete weld from both sides.

The leading edge of the radius pieces was left long, and the same tangent point and batten process was used to finish the cut. The hull side proceeded in much the same manner up to the forward area of plating. The forward area of the hull presented areas where the plating was both curved and twisted. My first attempt at plating this area did not go well. On my first attempt, I used a piece of plate that was too wide so it formed a big wrinkle in the middle which I could not get out. The second attempt using narrower pieces went better but not perfectly. As the plate was eased into position, a wrinkle started. I made a cut along the high point of the wrinkle and let the plate overlap as it was worked into place, trimmed the excess off and welded up the cut, treating it as I would a joint. The leading edge of the top sheet was marked out at the deck line at each frame and the batten was again used to get a nice fair line at the deck intersection.

Welding of the inside of the plate joints came next, starting near the middle of the boat, using short welds about two inches long spaced twelve inches or more apart and letting the weld cool before continuing the weld to prevent distortion. The next weld was started away from the existing weld and finished with a small overlap. Temporary stiff backs with a half circle cutout over the joint were tack welded across the joints on the outside to keep the joint flat while welding. After the inside of the joints were welded, the plating was welded to the scantlings and frames with stitch welds at alternating sides of the framing to prevent the plating from pulling into a visible bump again, starting at the middle of the boat. The tankage had to be welded continuously in short welds on both sides to make them liquid tight. The continuous welds showed an obvious ridge line on the outside of the hull. I opted to not grind the outside of this weld flat because it would thin the plating significantly. All the plate seam welds were air tested before the outside of the plating was welded.

The fuel and water tanks, which are integral, were completed at this time. The athwartship parts had been welded into the frames as part of the frame making process. The remaining parts were fit and welded with access to the inside made through a large access hole in the tank top which has a steel plate bolt-down cover set aside for later installation. Fittings for fuel and water plumbing and venting were made from steel threaded couplings welded into the tank plating. Baffles were fit into the tanks dividing them into quarters. The baffles were designed to be bolted to welded tabs so painting and repairs, if necessary, could be done with the baffles removed. Most of the decks were installed at this point, leaving the inboard edge long so they could be cut later when the hull was turned, pulling dimensions off of the cabin that had been assembled temporarily at the Calumet shop.

Most of the overhead welding was now completed. The end of putting out smoldering clothes was also almost done. Here is a tip, don't wear clothing made of synthetic material when welding.

A Little More Grinding and Hull Will Be Ready for Sandblasting and Primer

Synthetic material melts and can stick to your skin causing severe burns. Cotton clothing just smolders when ignited but can go undetected until your nose tells you to check your pants and shirt.

Welding of the outside of the hull involved back grinding the joint down to the inside weld with a cutoff wheel to insure complete penetration. The "short weld and skip" procedure was again used to prevent distortion. The outside welds were carefully ground down, taking care not to remove too much of the plating.

The tube for the bow thruster was the last major item before sandblasting and painting. Research for a bow thruster was a dilemma of sorts. Some makes were metric, and some were imperial in dimension. This had a direct bearing on the size of steel pipe to use for the tube. I decided on Key Power which required a schedule 40, eight-inch pipe for the thruster tube. After a lengthy study of the bow, I dared to cut the hole for the tube. Before the thruster tube was welded in place, the bow thruster was fitted to the tube while it was easy to access and then removed to be installed later. To make a fair radius at the intersection of the hull and the stern tube, I cut segments from steel handrail pipe turns I found in the shop. I was happy with the result.

Sandblasting was the most undesirable job of building the entire boat, even with good equipment. Inside of the tanks was particularly ugly to do, but it was an important phase of work. I would sandblast for part of the day and then coat that area before I quit that day.

I was fortunate to be able to contact the manager for the coating supplier for Marinette Marine. After doing a little research of my own and talking with the manager, we decided to use a Sherwin Williams product, Seaguard 5000 as the primer. It is a catalyzed coating that could be used on all steel surfaces including the inside of the fuel tanks. I purchased a ten-gallon kit of each gray and white so I could mix varying portions to get different tints for each coat.

I wanted to coat the inside of the potable water tank with a coating that would not give a taste or leach chemicals into the water. After much research I discovered a product called Brew Coat. The instructions said that the product was thick like molasses but do not thin it because it would add undesirable chemicals. The instructions were accurate, it was thick and sticky. It was mid-October before I was ready to coat the water tank, so after coating it I put a halogen work light under the tank to add a little heat overnight to speed the curing of the epoxy. I happened to pick the last day warm enough for the moths to come out at night. Morning inspection found the inside of the tank covered with moths stuck to the coating. A little gentle scraping took care of the issue.

With two coats of primer on inside and out, now it was time to start thinking about turning the hull over and building the roof structure over the boat. Good timing! Winter was setting in and preparations for the "big day" were underway which would be around Christmas time.

5 Right Side Up

Winter was closing in so concrete footings for the roof structure over the boat were a priority. Concrete piers were formed, poured, and covered to prevent freezing. Steel columns and beams were fabricated and painted. The truss and materials package had been ordered.

We were ready to turn the hull over. This was Christmas season 2018. I must have paced a hundred miles on the big event day. We had two extendable boom forklifts with long slings rigged with a chain fall on each sling to rotate the hull. When we lifted the boat, all the temporary support structure had to be removed. As we proceeded to rotate the hull, we reached a point where the boat was sliding in the slings, so plan B went into effect. We added another sling over the boat at midship, hooked to the gunnel and the other end to the bulldozer. We slowly pulled with the bulldozer as the hull slipped in the big slings until we reached the tipping point; then, things happened quickly. The boat completed the slipping in the slings on its own until right side up. What a rush! With some blocking under the keel to get her level and a few stanchions fabricated to hold her vertical, the job was completed.

In a few days we had completed the roof over the boat. Finally, I had a part of the project I had experience doing. Scaffolding was set up around the hull at a convenient working height and wood framing covered with reinforced plastic to enclose the boat was erected to keep out the winter weather.

The winter was spent working in the sheltered boat in nice weather and in the shop in not-so-nice weather. The deck framing and plating was completed, and the deck was marked out and cut using the mockup as a guide for dimensions. A three-inch vertical band was welded in place at the perimeter where the cabin would fit to make a watertight joint with the cabin side.

After the decks were completed, the last great challenge for the steel hull was started. This was forming and easing the four-inch half-pipe in place around the deck line. A lot of heat shrinking and tack welding and tugging with the coffin hoist brought the half pipe into place. The bow turn was made by first making a mock-up out of PVC pipe with darts cut out of the edges to use as a pattern. Both sides of the boat were worked at the same time to minimize the possibility of distorting the shape of the boat. With all the hours spent heating, welding, and adjusting, I only had one mishap—a tack weld broke loose, and the half-pipe sprang out, hitting me in the thigh. Thankfully, my cell phone was in my pocket which took the brunt of the force. Showing Kathy the banana-shaped phone was the painful part.

Fuel fill pipes were welded in place which extended from near the bottom of the side tanks at deepest point to a foot or so above the deck. This gave us the option of dipping the tanks with a marked stick to check the fuel level—it also gave the boat a little "work boat" feeling. The coaming was a fun part of the project. As the coaming shaped up, the hull came to life and finally looked complete. The coaming frames meant dreaming up another set of gigs to get them straight. The cleats and hawse-holes fabricated during winter were fit up and welded in place. Details such as steps and bow roller were finished. Shaping the half pipe and coaming required a lot of "eyeballing" to make them look fair.

The rudder shaft was made from 316 stainless steel at REL, a local shop. The half-inch steel plate rudder was welded to it. Duralon bearings were turned in our shop for the top and bottom rudder bearings. A packing gland was made to fit at the

A pause in the Hull Turning Operation – Time to Get Some Help from the Bulldozer

top of the rudder shaft tube and brackets were welded in for the steering hydraulic cylinder and autopilot. This completed the work on the stern.

The two-inch diameter propeller shaft was fabricated at Calumet Machine using 316 stainless steel. Two nonmetallic cutlass bearings were installed in the stern tube. Guidelines for bearing spacing came from one of my many boat building books. It was important to use nonmetallic bearings to prevent electrolysis between the shaft and the hull. The inboard end of the shaft was fitted with a Duramax dripless shaft seal which has nonmetallic mating surfaces to prevent electrolysis.

The keel cooler construction was an ongoing part of the hull and keel work. The keel cooler was partitioned off as the aft section of the box keel. The stern tube also ran through this section of the keel. Lubricating water for the stern tube required welding in small pipes from the keel sides into the stern tube near the forward end with water scoops added to the exterior of the keel. To circulate the coolant in the cooler tank, a cover plate was made on the forward end with a fitting near the top with a tube extending to the far end for the hot engine water to enter and a fitting near the bottom to supply cool water to the engine. The cooling system holds about forty gallons of coolant. Coolant is transferred between the engine through one-and-three-fourths-inch steel tubing and radiator hoses to take up movement. The engine coolant circulating pump moved the coolant—no other pump necessary.

Where I had planned to put the Aquadrive at the end of the propeller shaft, I realized, would not provide room to remove or install the aft bilge pump. I added a two-foot stub shaft to the prop shaft and secured it with a heavy-duty shaft coupling and installed the Aquadrive at the next frame forward. The mount for the Aquadrive had to be a substantial fitting because it had to withstand all the thrust of propelling the boat.

After the remaining steel parts were sandblasted and primed, the engine was fit up to the Twin Disc MC 5050SC transmission

Right Side Up – A Good Chance to See the Integral Tanks in the Hull

with a 2:1 ratio, trolling valve, and a PTO shaft to drive a small hydraulic pump. The engine beds had been installed as part of the hull fabrication but had to be modified a bit to match up with our fabricated engine mounts which were bolted to the engine and transmission.

It was here that lessons were learned about how easily stainless-steel bolt threads can be galled to make the nut impossibly stuck in places. This happened in a spot where access was very limited. An Aquadrive is a thrust bearing system with rubber mounts which isolates the thrust of the propeller shaft from the transmission and engine. Between the Aquadrive and the transmission, there is a CV shaft which had to be measured and ordered from Mac Boring, where I had purchased the Aqudrive system. Mac Boring guided me through the process of tweaking alignment of the transmission for minimum vibration. This set-up along with softer engine mounts, which were also part of the Aquadrive equipment, provides a quieter ride with less vibration. The Aquadrive system also allowed me to lower the engine and install the propeller shaft in a position closer to horizontal because the transmission did not have to be in a straight line with the propeller shaft. The near-horizontal prop shaft also minimizes "prop walk".

Our mechanic friend, Tom, would stop by in the evening now and then to sit on the staging, chat, give free expertise, and exercise his elbow. He would remind me to leave room for an engine with two more cylinders because the 4BT was not big enough, but Tom was into racing cars as a hobby.

In the early stages of mulling over what equipment would be needed for the boat, I decided with the use of LED lighting, a good solar panel system, and a large enough house battery bank, I could do without a generator. The refrigerator is the only big draw on the house battery when not underway.

The electrical system was not much more than a circuit breaker panel and a shore power receptacle. On the 120-volt side, an Isolation Transformer was selected to prevent electrolysis from stray current when using shore power. An inverter-charger was a must for charging batteries while on shore power and providing limited 120-volt power while underway. Three large AGM batteries were selected for the 12-volt house bank, and a conventional battery was selected for the start circuit. We had to find a location for the house electrical system that was central and consideration for this system's weight had to be put into the equation. More on electrical later.

The bow roller fabricated during the colder days was mounted through a hole cut into the coaming on the bow so it would be flush with the deck. We turned a Duralon roller at the tip to fit the three-eighths-inch anchor chain. The setup was fitted with a chain stopper and bolted to clips welded to the deck so the bow roller assembly could be removed. While access was still easy, the anchor windlass was fitted to the deck to align with the bow roller.

The chain locker bottom was installed using a piece of galvanized grating I found around the shop. The bottom was kept as low as possible to keep the weight on the 200 feet of chain closer to the waterline. A sperling pipe was fabricated to guide the chain from the windlass to the center of the chain locker. The sperling pipe ran through a shelf higher up in the chain locker, which could be used for light storage of such items as spare lines.

ABYC Standards, according to the size of my boat, required a total bilge pump capacity of 6,000 GPH. I placed two 1,000 GPH pumps in the keel and a 4,000 GPH pump in the lowest point of the hull. The keel pumps were mounted on stainless steel brackets with a vertical leg, bolted to the keel plate's top edge, making it easy to pull the pump for servicing when necessary.

The derrick to lift the dingy was at the end of the list of equipment to be fitted out. The derrick hardware was fabricated from stainless steel odds and ends and skillfully TIG welded by my nephew Roger. We started thinking that a simple block and tackle would do the job of raising and lowering the dinghy, but as we worked our way through the effort in our mind, we realized that with this method we didn't have enough hands. An electric windlass controlled by a foot switch was added for hoisting—this helped but did not solve all the problems. A worm gear drive boat trailer winch was installed to adjust the boom—this was a great improvement.

Now we were moving on to the cabin work, and this meant rounding up another set of tools.

6 Assembling and Trimming the Cabin

Many hours were spent sketching and pondering how best to lay out the interior. I decided to shrink the width of the side decks a few inches from what the plans showed to give us a little more room in the trunk cabin. If you can, imagine fitting all the essentials of a house into a space of about eight feet by twenty-five feet. There were always little things such as making beds long enough, cramming all three essential parts of a bathroom into a space about four feet by five feet, laying out the kitchen to include all the appliances including a microwave oven. Then there was the washer/dryer that sat collecting dust. It was measured and measured again as I struggled to find a home for it.

A big challenge was designing a plumbing system to include a water pump, filter, and expansion tank; finding a spot for the propane in-line water heater that would conform to ABYC standard, estimating water usage to size wastewater tanks (black and gray water), and shoehorning the tanks into the engine room. Install the plumbing so it can be winterized at the close of the boating season.

A list of additional critical dimensions and locations of many items had to be considered. The opening dimensions of the doors were checked to ensure appliances and parts could fit through. The proportions of the ship ladder-type steps and headroom at the steps were also checked. The location of access hatches and removable panels for electrical wiring modifications and equipment repairs was laid out.

The floor plan was critical, but deciding on the lines of the superstructure was equally important. Many items had to be incorporated into the exterior design, such as where to cradle the dingy, how to place and rig the lifting derrick for the dingy, termination of the exhaust system, and where to place the radar, horn, and spotlight. I also wanted to leave a free area for solar panels on the pilot house top.

Finally, the framing started. It was winter 2018. The framing was red oak from our woods—most of which has sat for years drying. The cabin overhead beams were shaped from the curved log we brought home a few years before. The framing was fit up to the mock-up of the deck detail I had built the winter before. Every part was angles and curves. Parts had to be made so they would fit out of the shop door when taken apart. Laying out the visor above the pilothouse window was a refresher course in high school geometry. The shape is a strip from the side of a cone. It fit first try!

The March sun warmed our plastic enclosure so work could resume on the decks and combing. After the deck and combing were sandblasted and primed and the running gear was put in place, the cabin parts were brought down from our Calumet shop, and assembly started. There weren't too many surprises as the framing fit up to the steel work very well. The big surprise was how close the pilothouse top came to the bottom of the trusses of the enclosure—this made for a quick, reassuring review of the overall height. We were still less than fourteen feet—the clearance height we needed for transporting down the road.

As the cabin assembly progressed, many more sketches, patterns, and cardboard mock-ups were made. The interior had to function, and all items had to fit. At every step, some details had to be thought out. Seeing the plan in 3D made for some fine-tuning of the plans.

The cabin shell was sheathed with half-inch marine plywood and covered with fiberglass cloth and epoxy resin. The exposed plywood edges were wrapped with fiberglass seam tape for extra

Andy, the guy that gave us the Junky RV,
Looks over the finished Cabin Woodwork with Joe

A Little Visitor to Provide a Welcome Break from Sheathing the Cabin

protection. The band around the cabin top was made from a thin plywood pattern transferred to a clear cedar board. The board was re-sawed in thirds and bent around the cabin top, bedded in epoxy resin with a layer of fiberglass cloth between each layer and a covering of fiberglass and resin.

We needed to add support to transfer the weight of the derrick and dingy to the hull structure, so we installed a post extending from the cabin top to the cabin sole. Below the cabin sole, a steel support was placed during hull construction. A small oak tree trunk sanded up and varnished was put in place. It is quite a conversation piece.

Going back to the donated RV, I thought of reusing windows from it but realized they were only one-eighth-inch tempered glass. With a little coaxing from Joe, I decided to bite the bullet and get real marine-grade windows. Diamond Sea Glaze windows were selected with three-eighths-inch laminated glass for the Pilot House and quarter-inch gray shaded laminated glass for the cabin. Smart move.

PEX tubing and PVC pipe caps were installed for the 12-volt electrical raceway systems, and larger PVC conduit raceways were placed where several wiring circuits were anticipated. We were ready for spray foam insulation using closed-cell foam.

Spray foam was applied to all exterior surfaces down to the waterline. I would guess that one-third of what was applied was shaved off because the space to receive foam was not very deep. What a mess to clean up! Small bits of foam were chased around forever—even after a good vacuum job.

Hidden behind all the work that shows, there is a system we installed that makes the difference between a fishing tub and a yacht—it is sound attenuation. Sound travels through hard structures and finds its way through every crack and hole. A sound attenuating system should be made of materials of varying densities to absorb sounds of differing frequencies. For example, the engine room overhead has mineral wool covered by perforated corrugated aluminum below the plywood deck and heavy vinyl sound absorbing sheeting covered by vinyl plank flooring on top of the plywood. Vertical engine room surfaces have self-sticking sound panels adhered to the plywood like you would find inside of a car door or trunk or your kitchen sink. Cracks and holes are calked or stuffed with insulation. The exhaust in the engine room is wrapped with high-temperature tape and the turbocharger has a custom blanket enclosure.

Laying out the position for gauges, equipment switches, engine controls, and many other gadgets was a real study of where to place them. Some items had to be accessible by everybody in the pilothouse. Some had to be operated with the right hand while steering with the left hand. Critical switches for wipers and bilge pumps needed to be placed so the order could be easily understood, and the screen for navigation instruments and the autopilot had to be in easy view of the captain.

The wainscot lining and cherry paneling installation proved to be very time-consuming once I found out that no two pieces were the same. The radius corner window trim was a challenge but after a few tries, MANY HOURS, and making special cutting gigs, it was a touch worth all the effort.

The exterior cabin doors were made of marine plywood with a backing of cherry frames with oak panels on the interior. The interior doors were oak framing with oak plywood inserts. The step threads came from the firewood pile. While splitting a few big oak blocks, I opened up beautiful quartered curly oak that I split as wedged and set to dry for use as the threads.

The cabin sole (floor to you landlubbers) was framed from Douglas fir two by threes and attached to the steel framing with stainless steel bolts with tabs welded to the hull framing to anchor support legs. The sole was five-eighths-inch fir plywood. All exposed wood surfaces were painted to minimize that musty boat smell. The sole had many hatches for access to equipment in the bilges. Each hatch had an aluminum molding around the outside and on the hatch itself to hold the heavy-duty vinyl flooring in place.

The aft deck was constructed from rectangular aluminum tube framing covered with a marine-grade aluminum quarter-inch diamond plate. The aluminum decking and framing were attached to the steel using stainless steel stove bolts with isolation washers and gaskets to minimize corrosion between the steel and the aluminum. The propane locker and stowage locker were also fabricated from the aluminum plate. Deck hatches were fitted with stainless steel hinges and recessed finger pulls.

My son Joe offered to build the galley cabinets. The counter-tops are made of IPE—a very dense hardwood from Brazil and the framing is oak. He did a first-class job.

Designing the helm area took some planning to organize all the functions. There were shift and throttle cables, cable for the trolling valve, helm pump for steering system and associated piping, defroster system, instrument backbone cables, Fireboy wiring, and a pile of wiring for all the gauges and switches at the helm. A raceway space was built between the forward pilot house windows to route electrical wiring and instrument cabling to the pilot house overhead. The wiring and cabling were routed in separated spaces to prevent electrical interference.

A few modifications were made to the interior of the boat after settling in. The original helm chair was a leather office chair—very comfortable, but it took up too much space, so we replaced it with a simple boat seat. The pilot house top bunk was lowered a few inches to make more space for the lucky top bunk person. AND a comfortable lounge chair was added to the cabin for the captain. Everything else on the interior is working out well.

7 Installing Systems

Installing systems was an ongoing project. A lot of the preliminary work had to be done before access was restricted. How and where equipment was placed took a lot of thought and planning, and remembering that one day, that part would have to be repaired was important.

The anchor windlass and the bow thruster have hydraulic motors. The long runs for the hydraulic system were installed as steel pipes with hydraulic hoses to make the final connection to accommodate movement and reduce noise. The hydraulic equipment is controlled by a valve bank designed for a snowplow with solenoid valves. Power was provided from a gear pump bolted on to the marine transmission and controlled by a pneumatic operated Logan clutch powered by air from the air horn compressor. A red light at the helm indicates when the pump is engaged.

A Kobelt steering system with helm pump and autopilot pump are plumbed with copper tubing with hydraulic hose at the autopilot pump and rudder cylinder. There is no reservoir on the steering system other than to fill the helm pump completely, which is at the high point of the system. The helm pump is a manual hydraulic pump operated by turning the steering wheel to transfer hydraulic oil from one side of the steering cylinder to the other to move the rudder.

The trolling valve is operated by a push-pull cable to adjust the flow of transmission oil which allows the transmission slip to

Dropping in the Engine and Transmission

Installing the Propeller and Rudder

slow the propeller shaft speed. A red light at the helm indicated when the valve is not completely closed. How simple is that?

The domestic water system has a storage tank with a capacity of about 100 gallons. A pressure pump and bladder tank give the system a relative even flow. An in-line propane RV water heater was installed using the enclosure as manufactured to prevent leaked propane from entering the cabin and can only be accessed from outside. The remainder of the system was typical residential parts until it came to the waste tankage and toilet. The toilet is a Vacuflush which requires a vacuum pump to operate it. The toilet is flushed directly into the black water tank, which can only be emptied by an evacuation pump at a marina. The gray water tank receives all other wastewater. Gray water can be held in the gray water tank, discharged overboard, or drained into the black water tank, depending on wastewater regulations where we are with the boat.

The washdown system starts with a steel pipe covered with a removal grate through the hull below the water line. The pipe terminates above the waterline and is routed through a ball valve and strainer before the pump. A hose bib is provided at the bow and in a locker at the stern.

The fuel supply system has a port and starboard tank with a capacity of about 110 gallons each. A day tank with a capacity of about forty-five gallons is built into the keel under the engine. All fuel that is transferred via the transfer pump goes through a large water separator filter. Fuel can be transferred from one tank to another for filling, ballasting the boat, or polishing bad fuel. The day tank is always filled with filtered fuel.

There were two options for building the exhaust system—a water-cooled system terminating at the transom with cooling water pumped into it after the exhaust manifold or a dry stack system like you see on many work boats terminating above the boat. The dry stack system has some benefits such as no possibility of water leaking into the boat and no exhaust billowing back into the aft deck. The exhaust piping we installed is all stainless steel, connected with v-band clamps. The exhaust diameter is three inches. Where the exhaust is routed through the cabin it is shielded with a five-inch exhaust pipe which allows ventilation air from the engine room to exit at the top along with the exhaust. In the false stack, the three-inch pipe stops short of the five-inch pipe, so the engine exhaust draws air from the air space along with the exhaust as it exits the stack.

The use of propane on a boat is very convenient. If the propane system is not installed according to ABYC Standards, it can be very dangerous—downright deadly. Propane is heavier than air, so any leak inside the boat can collect in the bilge, just waiting for the right conditions to explode. Unattended appliances, such as the water heater must be installed so any leaked propane will drain overboard. Attended appliances, such as the stove, cannot have a standing pilot light and must have a propane monitoring system with a solenoid shutoff valve in the propane locker. The monitoring switch must be readily accessible without reaching over the stove in case of fire, and the detector must be installed at the low point near the appliance. A sealed propane tank locker with a drain at the bottom to vent leaked propane overboard must be installed for tanks bigger than one pound. No fittings shall be used except in the propane locker and where the supply is connected to the appliance. This is just a brief description of the installation of a propane system as outlined by ABYC.

The electrical system just grew and grew. Soon after starting, the original order of marine grade wire and heat shrink connectors ran low. My good friend Jim Melchiore spent countless hours installing wiring and equipment and figuring out the best way to make the system work. The house battery bank and related high amperage wiring took a lot of head scratching and time to make first-class connections to that big wire. We added a sub-panel to the 12-volt system at the helm and several small fuse panels to protect the equipment in addition to the main panel. Being a steel hull boat, any circuit in the steel hull that could get

Jim Installing Twelve Volt Electrical System

Little Elmer on an Imaginary Cruise

wet or otherwise short out to the steel hull had both the positive and negative wire fused. We had to remember that all of the electrical, both 120 volts and 12 volts, had the grounds bonded together at only one point, which was the engine block, and must not be grounded to the hull at any point.

A semi-rigid solar panel system with the capacity to keep the refrigerator charged when the boat was not attended was selected. While researching solar panels, I came across an article about how the charging system from the solar panels would give bad information about the house battery status to the external voltage regulator on the alternator. A simple toggle switch to turn off the solar system when underway would have solved the problem, but who would remember to turn it off? We found a "normally closed" relay, which was wired into the solar panel circuit but powered by the engine alternator circuit, which shuts the solar system off once the engine is started.

A battery combiner was also installed. The combiner gives us the means of tying the house battery to the start battery if the start battery should go dead. A start solenoid powered through a momentary switch from the house battery completed the circuit for emergency starting.

Early on, we decided to install a Fireboy fire extinguishing system for the engine room. The theory is simple. When the fusible link on the fire suppressant tank gets hot enough, it discharges and puts the fire out. Ah, not quite that simple. The Fireboy system also included a panel that would automatically shut off the fuel to the engine, cut off the combustion air to the engine, cut off the intake air to the engine room, and stop the bilge blower. All these goodies prevented the fire suppressant from being sucked out of the engine room. The Fireboy panel also has a circuit to override the shutdown function so the bilge blower can be turned on to ventilate the engine room to make it safe to enter after the fire is extinguished. All we had to do was find the parts to complete the tasks and wire it all up. Tying in the fuel solenoid and the bilge blower was simple. We found a spring-loaded butterfly intake valve used on race cars for engine air intake. The engine room air intake was fitted with a six-inch diameter gasketed air damper and fitted it with a spring-loaded linear motor and a duplicate trigger mechanism, like the mechanism on the engine air intake.

Selecting an electronic system was more complex than just a cost decision. The system had to give us the information we needed in an easy-to-operate format. The Raymarine Axium Pro was selected as the heart of the system, with the Evolution system selected for the autopilot. The backbone system incorporated the radar, depth finder, heading sensor, and GPS. It is important to keep electronics cabling separated from electrical wiring. Wiring that provides power to electronics should be a twisted pair to minimize electrical interference to the electronics.

A marine radio with an AIS capable of receiving and sending (transceiver) was selected. The AIS needed to be registered and programmed to provide it with an individual identification code by the seller. We turned the radio on and like magic the position of *Kathy Ann* appeared on the Raymarine screen.

And, of course, there is the audible signaling device or horn as it is known. We did not select any horn. We installed a Kahenberg air horn system. There is no mistaking your presence when you toot.

Cold mornings and cold Lake Superior water splashing on the pilothouse can fog up the glass. Steering with one hand and stretching to reach with a towel to wipe the windshield with the other is an exercise that many boaters can relate to. A Jeep heater was installed with ducting to the pilothouse window bases and a small duct is run to the head. The engine cooling system is the source of heat for the heater. The heater is higher than the engine cooling system, so a plate heat exchanger with a small 12-volt pump was used to transfer the heated water to the defroster system.

The 110-volt electrical system was simple. Power for the microwave, lighting in the engine room, refrigerator power, and a few receptacles for power tools or the vacuum. The 110-volt source was either shore power or through the voltage inverter.

The 12-volt wiring is run through PEX tubing with two-inch PVC pipe caps fabricated for use as electrical boxes where access would not be possible after finish materials were installed, or other areas where access was limited. Spiral looming is used in areas such as the engine room to keep wiring orderly and easy to identify. Heat shrink connectors are used to prevent corrosion of connections. Heat shrink solder connections are used at critical connections. Sizing of the wire to minimize voltage drop is dependent on the length of the run FROM AND RETURN TO the battery and the calculated amperage for the device. Experience has put me in a tough spot in the past, when voltage drops below ten-and-a-half volts at the device—it shuts down.

We had a few minor bugs to work out of our systems, but the most puzzling quirk was the discrepancy of what the fuel gauges

read and what was in the tanks. We did not realize that the American system and the European system of sending units and accompanying gages are different. We have American sending units and European gauges which detect ohm values on opposite scales. The new gauges solved the problem.

Building a boat is a project that is never quite finished. Adding new gadgets, modifying existing functions, and the like go on forever.

8 Commissioning

As warm weather returned, the completion of *Kathy Ann* was in sight. I was crossing items off the top of the list faster than I was adding them to the bottom. As small projects were being completed, Julie Melchiori showed up with the last upholstered cushions and curtains. Her work set the interior off perfectly. The covers for the pilothouse windows were fitted with snaps. The helm was tidied up—everything was put in place. Big Jim did start up on all the appliances, equipment, and electronics. An Operators Manual was started with diagrams of all the electrical panels and sub-panels with fuse sizes and circuit descriptions. All the equipment information was rounded up to be put into a file.

It was time to start the engine. It started after the fuel system was bled and ran perfectly, but at the idle speed of 600 RPMs, it shook an undesirable amount. A little reading told us to increase the idle speed, so I gradually increased it until it smoothed out at about 750 RPMs. After a time, the engine began to miss. We could not find the problem after trying everything we could think of. The missing got worse, and finally, the engine would not start. We purchased a new set of injectors and installed them. Success... so we thought. The engine, again, started to miss and, after a while, quit running. We purchased a rebuilt injector pump and another set of injectors and installed them. The engine started after bleeding and ran perfectly. Hindsight gave us the answer to the problem. The engine had sat for at least 15 years with fuel in the system. When the engine was tested, fresh fuel filled the system and then sat for about two years loosening junk that had collected in the old injector pump. The junk then found its way into the injectors when it was started up.

My son Joe talked me into installing a Webasto diesel air furnace, the type that are often installed in semi-truck cabs. The unit came with all the necessary parts for complete installation, including a diesel fuel pump, ducts, exhaust tubing, and intake air tubing. By now the engine room was getting complex. There was one convenient spot left, that was over the top of the gray water tank. That little furnace is quite a piece of engineering. After operating the furnace for a while, we checked the engine room and found the materials near the exhaust to be hotter than we would like so we added an aluminum dryer vent duct to provide an air space and another layer of high temperature insulation over the exhaust tube.

A collection of tools, spare parts, spare filters, and engine fluids had accumulated by this time. Finding a convenient place for everything was something I felt would work itself out as we ran the boat. There was one empty space left in the engine room—the top of the 10-gallon hydraulic oil tank which made a perfect place to sit while working in the engine room.

The trailer had yet to be built because the wheels and spindles were back ordered forever. The trailer was designed to keep the boat as close to the ground as possible. We needed to keep the overall height at fourteen feet while on the road. The boat would sit on the trailer with a forward slope to keep it low. Rather than have axles as you normally see on a trailer, I decided to put the wheels in pairs on each side of the perimeter beam with cross pieces slung below the perimeter beam. When the wheels finally showed up my nephew Roger, the real welder, worked with me to measure up and fabricate parts which were slid into place under the boat and raised up to fit the hull. The wheels were put on last; the blocking was removed and the whole works was lowered to the ground. We had to do a little excavating and reshaping of the work area to ensure that we could clear the beam at the edge of the roof over the boat once we started to move the boat. We had a pintle hitch mounted on a fork of the forklift to do the close quarters move. We moved *Kathy Ann* to level ground in front of the shop to finish up a few items and load up items which needed to be installed once moved to the boat launch.

We did a reconnaissance run to the boat launch to check clearance under wires crossing the road to be sure they were all fourteen feet high or more. The big Farmall tractor was the chosen piece of equipment to pull *Kathy Ann* to the boat launch because of its ability to adjust the height of the hitch with its hydraulic drawbar. The move started at first light on Saturday morning with one truck with caution lights on in front and another at the rear. The move went without a hitch. The forklift was brought to the boat launch to hook up to the tractor for the launch "just in case". It took several hours to install the items on the pilothouse top and the exhaust stack which we thought would not clear the wires. The anchor chain was flaked out on a trailer bed, marked at twenty-five-foot intervals with colored zip ties and fed into the chain locker.

While preparing for the launch there was quite a parade of spectators to see the boat as word must have spread about the trip through Lake Linden in the early Saturday hours. The launch went smoothly, and we fired up the engine and checked over the boat for problems before we took it on our first short run. Finally, we were underway, and the excitement diminished. I looked around and I realized we had a few extra crew members aboard—a total of thirteen people. Four of us were the launch party, three of the additional crew were the local mechanics, and the others were trusting souls who really wanted to be on the first trip.

Our only problem was a leak where the bow thruster was bolted to the tunnel. We decided to fix it while in the water rather than hauling it out again. We nosed the boat into shore and lifted the bow with the forklift, so the thruster was clear of the water. My willing nephew, Roger, took his boots off and went swimming. We handed him tools to disassemble the thruster, reapplied more bedding, and bolted it back into place without dropping anything. The next day we ran the *Kathy Ann* to the H&Y Marina in Hancock where we completed a few work items to get her ready for the trip to her home port, Agate Harbor.

Joe and I made the trip to Agate Harbor on a beautiful late summer day. The color of Lake Superior is a blue like no other lake. We were running quartered into a two-to-three-foot sea most of the way. The boat rode the sea well as the plumb bow broke through the waves. There was some occasional spray on deck and a few occasions where we had to test the wipers. We had an opportunity to try out Our Raymarine autopilot by plotting points ahead and letting *Kathy Ann* steer herself.

We tied up at Bob Carlton's home. This dock was a project I had built about twenty-five years earlier, a dock I built to withstand the movement of the spring ice. *Kathy Ann* stuck out about eight feet past the end of the dock, so we were concerned about the west wind pushing against her. We did some dock repair and reinforcing and added a few extra cleats and dock lines.

Early that summer I realized what had been created was now worth a good piece of change, so it was time to inquire about insurance. I found out in short order that getting a home-built boat insured would not be easy. Having spent years looking at how home-built projects were sometimes put together I could understand. Before an insurance company would consider insuring this boat, I needed to get a report from an ABYC certified marine surveyor. We contacted a surveyor from Bayfield Wisconsin, Phil Peterson, who was familiar with our area, and he agreed to make the trip to look over the boat while on the hard. Overall, the inspection went very well. Knowing that we had built to ABYC Standards was a good start for his day. There were a few items which needed a little tweaking and a few minor items which we had not yet completed. Based on Phil's report, the

***Kathy Ann* Set to Transit to Her Home Port – Agate Harbor**

insurer gave the approval for launch and travel to our home port only.

As required by the insurer, the marine surveyor had to come back to do sea trials. After our mid-August launch and our five-hour trip to Agate Harbor, we had a little time to correct and complete the items on Phil's report. The marine surveyor came, and we took *Kathy Ann* out to sea (Lake Superior) for sea trials. All went well on both the time runs against the sea and with the sea, gave the autopilot a workout, and then took her into the harbor and did our start and stop maneuvers. Engine room temperatures were checked a few times and the engine mounts were observed during start and stop maneuvers. Two problems were noted on sea trials: the swim platform was mounted a few inches too low, which caused water to splash back into the stern deck scupper, and the 110-volt grounding needed to be rerouted. The grounding problem took a study of diagrams from each part of the 110-volt electrical system to understand how different pieces of the 110-volt electrical system should have been connected together. The solution to the electrical grounding problem was to simply run the ground as you would in the electric service in your house up to the isolation transformer.

My family put together a boat christening ceremony at the Eagle Harbor Marina for *Kathy Ann* in late August. There was a wonderful luncheon complete with speeches. I did not realize the magnitude of my accomplishment until I saw how many people were in attendance. Perhaps it was because I always had confidence that it could be built. What a great day it was.

The boating season was near its end for 2021 so the swim platform work and grounding issue would have to wait until spring. The *Kathy Ann* was brought back to H&Y marina late in the fall to be hauled and winterized, last boat to be hauled out to be more specific. We decided to shrink wrap the boat ourselves, so we bought the shrink wrap plastic and the equipment to go with it. When we were finished it showed that we needed a little

Buzz's Family Support Crew at the Christening at Eagle Harbor

more practice at building the frame and shrink wrapping, but it was good for the winter.

We drive through the marina now and then to check the boat covering job. Winter had been long, very cold, and a lot of snow. A new list was started for items that had to be done and items we would like to modify and add. After years of boating, you realize that a boat project is never quite finished.

Reviewing the boat plans, I found that, indeed, the swim platform was too low by about five inches. We cut the swim platform loose and welded it in place at the proper height. The paint job was touched up, and a plastic cover with a float ball for a backwash valve was installed over the scupper as an added precaution. With Phil Peterson's approval, the insurer granted us insurance.

2022 found us rethinking the hydraulic system. The pump selected and the way it was installed was not what we expected. A larger capacity pump was needed to drive the bow thruster and windlass satisfactorily, so we did a lot of research. A clutch operated by compressed air from the air horn system was purchased along with a simple gear pump with a flow rate about double that of the original. Problem solved—almost.

Work on our Agate Harbor cottage occupied most of my summer, but we did manage to have a wonderful trip to Isle Royale with five adults and five kids. The trip home after three days found kids napping in every corner.

On a late summer excursion, the engine overheated. We spent a lot of time thinking about what caused it to overheat. The cause was narrowed down to two possible items—the additional load on the engine because the hydraulic pump remained engaged, or the coolant header tank with a volume of about one-and-a-half gallons was too small for the forty gallons of coolant in the keel cooler system. Sometimes, the clutch would not disengage, so we retraced our steps. After a lot of disassembly, checking, and rechecking, it was discovered that condensate from the compressor was working its way into the clutch. To solve the problem, a water separator and drip leg were added to the air system.

The hydraulic system modifications had been working properly so it was moved to the bottom of the list. The volume of the header tank was doubled, and a high-level and low-level sight glass were installed in the tank. As a precaution, we pulled the engine head to see if the overheating had caused any damage. There were bad scuff marks in one cylinder, so the dilemma was to repair the engine we had or get a remanufactured engine to replace it. It was decided to get a remanufactured engine, not knowing what else could be damaged by overheating.

When the pilot house sole framework was installed, I made an opening just for such an operation—but not expecting it to be this soon. The original engine had to be stripped to the bare block to fit through the opening in the deck. The parts removed had to be installed on the new engine anyway. By now, the only piece with moving parts that had not been replaced was the turbocharger, so the turbo was replaced too.

The new engine was a direct replacement for the original, so the installation went well. This work gave us a chance to reroute a few wires and modify the pulley and belt arrangement for the house alternator. The new engine ran flawlessly.

As work on the boat progressed, spare parts were set aside for critical mechanical components. Everything from spare belts to injector pump, at least one set of filters, and lubricants are with the boat. Spare fuses, electrical connectors, and an electrical tester plus many more odds and ends are also packed away for "just in case".

9 Reflections

The love for boating and the outdoors was always there. From the first time Dad fixed me up with an alder trig with a piece of line, a hook and worm to catch little speckled trout in the creek that ran by the barn.

My first real boating adventure was a trip to Isle Royale with a group of 4-H kids aboard the *Isle Royale Queen II*, which was brand new that year. This was 1960. She rode like a cork. The infrastructure at Snug Harbor was in the process of being built, the *Ranger III* was just a few years old. The families of baby boomers were everywhere celebrating the experience of camping on Isle Royale. The grounds of the campsite were worn to bare dirt. The trip home was like being inside a washing machine with gale-force winds on the beam. I was one of the few not to turn green.

After completing college, my uncle Jack invited me to Isle Royale as part of a family trip. We rented fourteen-foot boats from NPS Concessions and brought our own motors. We fished and camped for about two weeks. During this trip we experienced a seiche that swamped our boat and caused the pulpwood deadheads that had drifted there from Canada to bob on end at the water surface. We also towed in a guy with a twelve-foot Styrofoam sailboat that did not know how to tack into the wind.

Over the next several years I was invited along on fishing trips to the island with my uncle Carl. Those days, navigation equipment consisted of a plywood box with a barometer, a good compass, a logbook, and a chart. The wristwatch completed the list of equipment. This was a twenty-four-foot inboard. There were also many other trips to Isle Royale with my friend Paul Pichette. Many more fishing adventures were made with our own boats.

I have built hardtops for boats for a few friends. I built a plywood dingy which we carried on the Chris Craft to Isle Royale a few times. But the jewel is working with my father on his traditional cedar-canvas canoes, Klondike Canoes. Elmer kept a log of his canoe production. He had built eighty canoes before he was unable to work in his shop. He and my mother delivered canoes from coast to coast as part of a paid-for vacation.

Back to the plywood dingy. We had taken the dingy along on a boating excursion to Isle Royale. A second boat in our party was owned by Gordy. Gordy had invited along his handyman Paul. We pulled into Birch Island at the entrance to McCargo's Cove for the night. Paul and I decided to try our luck at Northern Pike fishing using the dingy. We rowed into Brady Cove and made a few casts. We were knees to knees in the little dingy and dared not to make any unnecessary motion for fear of dumping the dingy over. Paul got a strike. It was a monster pike. He fought the fish as it pulled us around the cove. The big pike made a pass behind Paul. Paul had his rod over his head with the fish behind him and accidentally knocked his glasses off and into the water. He managed to keep the big fish on and slowly worked it closer to the dinghy. The big pike lunged for the bottom and snapped the end section of his rod off. We managed to get the monster into the landing net and into the dinghy. That was enough for that fishing excursion. On returning home, Paul pickled the pike and handed it out to us all. If you ever catch a pike wearing glasses in Brady Cove, give them back to Paul.

For you who are new to boating, remember, someday all that fancy stuff you install in your boat will not work. Learn how to navigate the old-fashioned way—using a compass, watch, logbook, and a means of determining your speed. This is called dead reckoning. Write notes about each passage in your log. Make a sketch of features for difficult harbor entrances with buoys and compass headings.

See you on the high seas.

About the Author

Robert Johnson is a lifelong craftsman, builder, and resident of Michigan's Upper Peninsula. With decades of experience in residential and commercial construction, he combined his skills and passion for the Great Lakes to undertake the ambitious project of building *Kathy Ann*, a steel-hulled trawler designed for Lake Superior's demanding conditions. Drawing inspiration from his Finnish and Swedish heritage, Robert's work reflects a deep respect for tradition, utility, and craftsmanship. When he's not on the water, he enjoys restoring historic structures, spending time with his family, and sharing his knowledge of building and boating with future generations.

Acknowledgements

I appreciate the time and knowledge others have contributed to the successful completion of the *Kathy Ann*. Without your help, the road to completion would have been a lot longer and the finished project not of the professional quality it is. I may not name a few of you but it is by no means intentional:

Jim Melchiori	Electrical design and installation
Mark Tuovala	Mechanical advice and sharing of vital equipment
Joe Johnson	Cabinets and lots of advice
Roger Johnson	TIG welding stainless steel and aluminum parts
Robbie Johnson	Lathe work on Duralon bushings and special parts
REL, Inc.	Rudder shaft fabrication
Calumet Machine	Propeller shaft fabrication
Universal Metal Works	Fabrication of bow plating
Dudley Dix	Designer of the *Echo 38*

Glossary for Landlubbers

Athwartship – Across the boat, side to side

Chine – Where the side of the hull meets the bottom

Coaming – Vertical projection above the deck to prevent water from entering boat.

Day Tank – On *Kathy Ann*, a fuel tank with a capacity to run the engine for about twelve hours. This tank is filled through a filtering system to ensure clean fuel for the engine.

Derrick – On *Kathy Ann*, it is the mast and boom which resembles the rigging on a sailboat for hoisting the dingy to the top of the cabin. Sailing ships used their rigging in the same fashion to hoist the anchor to the rail and also load heavy cargo.

Dog – In our application, a dog is a scrap of steel welded to one side of a plating joint so a wedge can bring the plates into alignment.

Fish Tug – Fish Tugs were older steam powered tugboats that were used for commercial fishing on the Great Lakes because of their seaworthiness. Enclosures were built on the decks to protect the fishermen n bad weather. This idea has evolved into what we see as working fishing boats today.

Flake Out – Laying out rope, or in our case chain on the deck so it could be fed out without tangling.

Hawsehole – a hole in the hull through which a mooring line (anchor rope or chain) passes.

Integral Tanks – Tanks that are built as part of the structure of the boat. Diesel fuel is classified and flammable (as opposed to gasoline which is classified as explosive) so regulations permit diesel to be stored in integral tanks made of fuel proof materials. This saves a lot of space in a boat

Limber Holes – Openings cut in the frames and scantlings of the hull to prevent water from accumulating at low points.

Hull Speed – A displacement hull boat such as *Kathy Ann* can only efficiently travel at speed it takes for the bow wave to travel the length of the boat. Any faster and the boat in effect starts to go up-hill. The longer the boat, the faster it can go.

Long Period Waves – The period of waves is measured from crest to crest. The greater the distance a wave travels (fetch), the longer the period becomes.

Loran C – A forerunner of GPS. The system has a series of antennas with differing frequencies. The position of your target point was a measurement of the time it took the radio wave to reach the target point from two antennas. Think of it as having two archery targets close to each other and measuring out from each center a given distance – where the circles intersect is the target point. This point was plotted from a chart with Loran circles overlaying it.

On the Hard – When the boat is on land such as in a boat yord, it is said to be "on the hard".

Packing Gland – A mechanical device made to prevent water or other liquids from leaking around a turning shat or a sliding rod. Impregnated flax is often used as a wrapping around the shaft or rod which is compressed by the mechanical gland.

Prop Walk – A sideways movement of the stern of an inboard propelled boat at slow speeds cause by the downward angle of the propeller shaft. Understanding how prop walk affects your boat can be used to your advantage, not understanding pop walk can get you in a lot of trouble.

RIB – An inflatable boat with a ridged bottom.

Running Gear – The engine, transmission, propeller shaft, and the propeller installed as a system.

Rudder Shoe – A boat such as *Kathy Ann* had a keel which is deeper than the propeller. A shoe is extended from the keel, under the propeller to capture a bearing at the bottom of the rudder post. This stabilized the rudder and also protects the propeller.

Scantlings – longitudinal steel members fit into notches in the frames which define the shape of the hull

Seiche – Similar to when you blow on a hot cup of coffee and the force spills it over the side, high winds on Lake Superior can cause a buildup of water on one end and receding water on the other, also known as seiche.

Sistered Rib – An additional rib or part of a rib placed along side of a broken rib in wooden boat repair.

Spurling Pipe – A pipe or tube installed to guide the anchor rode (rope or chain) from the windlass to the chain locker.

Trolling Valve – An adjustable valve in the transmission which allows some of the transmission oil to bypass the clutch allowing it to slip so the boat can travel at a slower speed for trolling or lifting nets.

Turn of the Bilge – A rounded chine as opposed to a sharp shine which enables to boat to ride a cross sea safer.

www.ingramcontent.com/pod-product-compliance
Lightning Source LLC
LaVergne TN
LVHW080530210926
848481LV00019B/1455